Cambridge Elements

Elements in Geochemical Tracers in Earth System Science
edited by
Timothy Lyons
University of California
Alexandra Turchyn
University of Cambridge
Chris Reinhard
Georgia Institute of Technology

THE IRON SPECIATION PALEOREDOX PROXY

Simon W. Poulton
University of Leeds

CAMBRIDGE
UNIVERSITY PRESS

CAMBRIDGE
UNIVERSITY PRESS

University Printing House, Cambridge CB2 8BS, United Kingdom

One Liberty Plaza, 20th Floor, New York, NY 10006, USA

477 Williamstown Road, Port Melbourne, VIC 3207, Australia

314–321, 3rd Floor, Plot 3, Splendor Forum, Jasola District Centre, New Delhi – 110025, India

79 Anson Road, #06–04/06, Singapore 079906

Cambridge University Press is part of the University of Cambridge.

It furthers the University's mission by disseminating knowledge in the pursuit of education, learning, and research at the highest international levels of excellence.

www.cambridge.org
Information on this title: www.cambridge.org/9781108810685
DOI: 10.1017/9781108847148

First published 2021

A catalogue record for this publication is available from the British Library.

ISBN 978-1-108-81068-5 Paperback
ISSN 2515-7027 (online)
ISSN 2515-6454 (print)

The Iron Speciation Paleoredox Proxy

Elements in Geochemical Tracers in Earth System Science

DOI: 10.1017/9781108847148
First published online: February 2021

Simon W. Poulton
University of Leeds

Author for correspondence: Simon W. Poulton, s.poulton@leeds.ac.uk

Abstract: In one form or another, iron speciation has had a long history as a paleoredox proxy. The technique has been refined considerably over the years, and the most recent scheme is unique in its potential to distinguish three major oceanic redox states – oxygenated, ferruginous and euxinic. This Element covers the theory behind the proxy, methods involved in applying the technique and potential complications in interpreting Fe-speciation data. A series of case studies are also provided, which highlight how more advanced consideration of the data, often in concert with other techniques, can provide unprecedented insight into the redox state of ancient oceans.

Keywords: Fe speciation, paleoredox, ferruginous, euxinic, anoxia

ISBNs: 9781108810685 (PB), 9781108847148 (OC)
ISSNs: 2515-7027 (online), 2515-6454 (print)

Contents

1 Introduction

The use of geochemical techniques for reconstructing ancient water-column redox conditions has been a key aspect of paleoenvironmental research for more than 50 years. Paleoredox proxies based on Fe-S systematics have been at the forefront of such studies from the earliest days, with some of the first calibrated redox proxies being based on the degree of pyritisation (DOP) of iron minerals (Raiswell et al., 1988). While the techniques involved in utilising Fe-S systematics as a paleoredox tool have evolved as understanding of the (bio)geochemical behaviour of iron and sulfur during transport, deposition and early diagenesis has progressed, the Fe-S system has remained a key component of the geochemical paleoredox toolbox. Indeed, although there are numerous potential obstacles involved in the successful application of Fe-S systematics (which is the case for every redox proxy), and while it is increasingly clear that a multi-proxy approach should be favoured, the Fe-S system remains the most widely calibrated and well-understood technique for evaluating local water-column redox conditions, with several unique aspects that particularly enhance its utility.

Although from a strict geochemical standpoint the terminology might be considered incorrect, the term 'Fe speciation' has become synonymous with the Fe-S paleoredox tool. A brief history of Fe speciation is provided here, and for more detailed historical insight, the reader is directed to extensive reviews of the proxy (Raiswell and Canfield, 2012; Raiswell et al., 2018). Iron speciation was initially used to identify controls on the formation of sedimentary pyrite (Fe_{py}), leading to the development of the DOP parameter:

$$DOP = \frac{Fe_{py}}{Fe_{py} + \text{HCl soluble Fe}} \tag{1}$$

Subsequently, the DOP proxy was calibrated to distinguish aerobic, restricted and inhospitable bottom waters (Raiswell et al., 1988; Raiswell and Al-Biatty, 1989), where Fe_{py} was determined via the chromium reduction method (Canfield et al., 1986) and a 1 minute boiling HCl extraction was used to define a 'reactive' Fe pool (Fe_R).

A turning point came with a greater understanding of the kinetics by which different Fe mineral phases react with dissolved sulfide. This abiotic reaction ultimately results in the reductive dissolution of certain Fe minerals, with a variety of Fe (oxyhydr)oxide minerals being particularly prone to sulfide-promoted reductive dissolution on early diagenetic timescales (Canfield, 1989; Canfield et al., 1992; Poulton et al., 2004a). While the boiling HCl extraction successfully extracts Fe (oxyhydr)oxide minerals (Raiswell et al., 1994), it also extracts Fe from a variety of sheet-silicate minerals (termed Fe_{PRS}; poorly reactive silicates), which are only appreciably reactive towards dissolved sulfide on a million-year timescale (Raiswell and Canfield, 1996). This limits the sensitivity of the DOP proxy to different redox states, and as a result, a sodium dithionite solution was developed (Canfield, 1989; Raiswell et al., 1994) as a more suitable extractant of minerals such as Fe (oxyhydr)oxides (termed Fe_{ox}). Raiswell and Canfield (1998) then defined a 'highly reactive' Fe pool (Fe_{HR}) as the sum of Fe_{ox} and Fe_{py} (note that here, Fe_{py} refers to pyrite Fe plus Fe present as acid-volatile sulfides, Fe_{AVS}). However, recognising that important highly reactive minerals were not extracted by these techniques, Poulton and Canfield (2005) further refined the iron-speciation methodology. This resulted in the development of a sequential extraction procedure to additionally determine mixed ferrous-ferric oxide phases such as magnetite (Fe_{mag}) as well as iron present in carbonate phases (Fe_{carb}). An updated highly reactive Fe pool (henceforward termed Fe_{HR}) was then calculated as the sum of Fe_{carb}, Fe_{ox}, Fe_{mag} and Fe_{py}, and it is the behaviour of this operationally defined Fe pool in near-surface environments that forms the basis of the current Fe-speciation paleoredox proxy.

A particular strength of Fe speciation is the potential to distinguish three major redox states of ancient oceans – oxygenated, ferruginous (anoxic, containing dissolved ferrous Fe) and euxinic (anoxic, containing dissolved sulfide). An oxygenated water column from the perspective of Fe speciation comprises a range of oxygen levels, from fully oxic to dysoxic, since the proxy cannot distinguish between different levels of oxygenation (see Raiswell and Canfield, 1998). In this regard, additional techniques are recommended for distinguishing oxic and dysoxic water column conditions, such as Re/Mo ratios (e.g., Kendall et al., 2010) or rare earth element systematics (e.g., Tostevin et al., 2016). However, a particular feature of Fe speciation is the possibility to distinguish oxygenated settings from both ferruginous and euxinic environments. Indeed, Fe speciation is the only paleoredox tool that has the potential, in isolation, to distinguish the anoxic ferruginous and euxinic redox states. As a direct result of this unique attribute, our understanding of the major significance of ferruginous conditions throughout huge swathes of Earth's history has been revolutionised over the last decade (e.g., Canfield et al., 2008; Johnston et al., 2010; Poulton et al., 2010; Planavsky et al., 2011; Poulton and Canfield, 2011; Guilbaud et al., 2015).

Here, we focus on the current state of the art with respect to Fe speciation, including a general overview of how the proxy works, details of the latest methodology that should be applied, and consideration of potential complications when applying Fe speciation as a paleoredox tool. In this last regard, potential ways in which some common difficulties may be overcome are discussed via case studies, and some prevalent misunderstandings of the proxy are also addressed.

2 Systematics of the Fe-Speciation Paleoredox Proxy

The Fe-speciation proxy is built upon a long history of studies of the (bio)geochemical behaviour of iron and sulfur during early diagenesis (e.g., Berner, 1970; 1984; Canfield, 1989; Canfield et al., 1992). While the proxy is primarily aimed at siliciclastic marine sediments, it is also possible, with care, to apply the scheme to some carbonate-rich lithologies providing they contain a significant siliciclastic or authigenic Fe component and have not experienced introduction of Fe_{HR} during deep burial dolomitization (e.g., Clarkson et al., 2014; Wood et al., 2015; Tostevin et al., 2016; Zhang et al., 2018). The utility of Fe speciation is predicated on anoxic enrichments in highly reactive iron relative to normal background concentrations found in marine sediments deposited under oxygenated water-column conditions. In an extensive global study of modern marine sediments deposited in oxic and dysoxic waters, Raiswell and Canfield (1998) found that Fe_{HR}/Fe_T ratios (Fe_{HR} is normalised to Fe_T to allow for variable dilution of the terrestrial input, for example by carbonate, organic matter and silica) were generally lower than 0.38. By contrast, modern euxinic settings commonly have Fe_{HR}/Fe_T ratios higher than 0.38 (Canfield et al., 1996; Raiswell and Canfield, 1998). Enrichments in Fe_{HR} in euxinic settings such as the Black Sea arise due to the water-column formation of Fe sulfides, where the additional dissolved or reactive particulate Fe is supplied via release from sediments on the shallower shelf, a mechanism that has been termed the 'shelf-to-basin Fe shuttle' (Anderson and Raiswell, 2004; Raiswell and Anderson, 2005; Lyons and Severmann, 2006). Settling of these Fe sulfides through the water column augments the terrestrial Fe_{HR} influx, giving diagnostic enrichments in modern euxinic basins (Canfield et al., 1996; Raiswell and Canfield, 1998; Lyons et al., 2003).

Under anoxic ferruginous water-column conditions, similar enrichments occur at sites where precipitation of dissolved Fe^{2+} is induced. In this case, however, the primary enrichment is in unsulfidised Fe_{HR}, and there is much debate and continued research on the mechanisms involved in Fe mineral precipitation under ferruginous conditions, in addition to the nature of the primary and secondary minerals that are formed and ultimately preserved in the sediment (e.g., Konhauser et al., 2002; Crowe et al., 2008; Zegeye et al., 2012; Rasmussen et al., 2015; Halevy et al., 2017; Li et al., 2017; Jiang and Tosca, 2019). While the shelf-to-basin Fe shuttle may also be a source of Fe in ferruginous basins, there has been a tendency to over-emphasize the role of this shuttle in studies of ancient anoxic settings, where water-column anoxia may have extended into the deep open ocean. For example, under the global ferruginous conditions that are now believed to have been a dominant feature of the ocean throughout much of Earth's history, upwelling of ferruginous deeper waters would have been a primary source for Fe_{HR} enrichments on the shelf (Poulton et al., 2010; Poulton and Canfield, 2011; Wood et al., 2015), with the Fe sourced either from hydrothermal input (e.g., Derry and Jacobsen, 1990) or via release from non-sulfidic sediment porewaters. Indeed, as noted by Poulton and Canfield (2011), under anoxic ferruginous conditions, deep-sea sediments may have been a source of Fe^{2+} to the water column rather than a sink, potentially making the identification of deep-ocean redox state particularly difficult.

Detailed calibration of Fe_{HR}/Fe_T ratios in ancient marine settings has confirmed the utility of the proxy for identifying ancient anoxic water-column conditions (Raiswell et al., 2001; Poulton and Raiswell, 2002; Clarkson et al., 2014). However, Poulton and Raiswell (2002) noted that Fe_{HR}/Fe_T ratios in oxic Phanerozoic settings are generally considerably lower (0.14 ± 0.08) than the 0.38 boundary that commonly distinguishes oxic and anoxic settings in modern oceans. This often arises due to transfer (and hence loss) of Fe_{HR} during clay mineral formation in anoxic non-sulfidic porewaters, which may have been a particularly extensive process in ancient, organic-rich ferruginous settings (e.g., Poulton et al., 2010; Doyle et al., 2018; see Section 4). In addition, anoxic water-column Fe_{HR} enrichments may be masked by rapid sedimentation (e.g., during turbidite deposition; Raiswell and Canfield, 1998). As a result, Poulton and Canfield (2011) introduced the concept of equivocal zones in the use of the Fe-speciation redox proxy (Figure 1). Thus, as a first assessment, Fe_{HR}/Fe_T <0.22 provides an indication of oxic water-column conditions, $Fe_{HR}/Fe_T >0.38$ indicates possible water-column anoxia, and intermediate Fe_{HR}/Fe_T ratios are considered equivocal.

Poulton and Raiswell (2002) suggested that during the Phanerozoic, the Fe cycle remained in a long-term steady state in terms of the global supply of Fe_{HR} to the ocean. However, this was based on limited available data for Phanerozoic rocks, and of particular potential significance for the Fe-speciation paleoredox proxy, it is possible to envisage a scenario whereby elevated chemical weathering on the continents could result in the supply of sediment with high Fe_{HR}/Fe_T ratios to the ocean. The present-day global river sediment Fe_{HR}/Fe_T ratio is 0.43 $\pm$ 0.03 (Poulton and Raiswell, 2002); however, even in the case of the Amazon river, which today supplies the largest flux of sediment to the ocean (Milliman and Meade, 1983), the highly weathered nature of the tropical catchment area only results in a suspended-load Fe_{HR}/Fe_T ratio of 0.47 (Poulton and Raiswell, 2002). Furthermore, Fe_{HR} is preferentially trapped in near-shore environments,

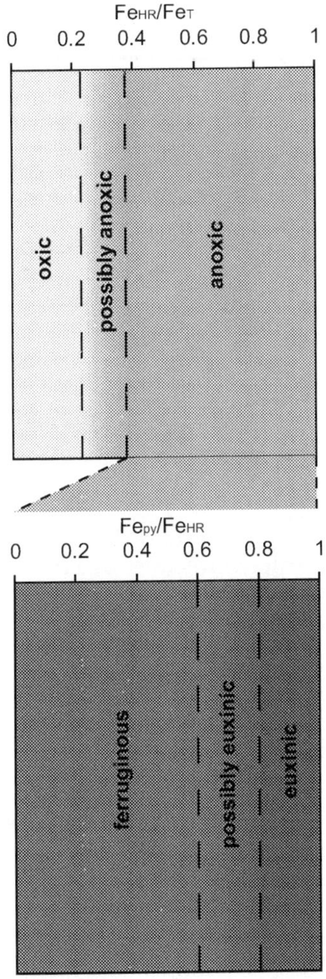

Figure 1 Conceptualisation of the iron-speciation parameters for evaluating ocean redox conditions (modified after Poulton and Canfield, 2011).

and thus the average Fe_{HR}/Fe_T ratio of modern marine sediments (0.26 ± 0.08) is significantly lower than the average for global rivers (Poulton and Raiswell, 2002). This suggests that significant terrestrially derived Fe_{HR} enrichment in ancient oxic marine sediments is likely to be the exception rather than the rule, and additional robust evidence would be required to significantly raise the upper Fe_{HR}/Fe_T oxic depositional threshold of 0.38.

As outlined earlier, a unique utility of Fe speciation is the ability to distinguish ferruginous and euxinic conditions. Poulton et al. (2004b) first utilised Fe_{py}/Fe_{HR} ratios for this purpose, for samples displaying clear evidence of deposition under anoxic water-column conditions (i.e., Fe_{HR}/Fe_T >0.38). The use of Fe_{py}/Fe_{HR} ratios stems from observations of the extent of sulfidation of the Fe_{HR} pool in independently verified euxinic settings. In particular, in euxinic Black Sea sediments, Fe_{py}/Fe_{HR} ratios commonly exceed 0.8 (Anderson and Raiswell, 2004), which, when combined with elevated Fe_{HR}/Fe_T ratios, is considered to provide a strong indication of water-column euxinia in ancient settings. However, it is particularly difficult to define a robust lower Fe_{py}/Fe_{HR} boundary for identifying euxinia. Indeed, März et al. (2008) found that a threshold of 0.7 was more appropriate for distinguishing euxinic and ferruginous conditions in sediments deposited during Cretaceous oceanic anoxic event 3, which led Poulton and Canfield (2011) to define an equivocal zone of 0.7–0.8. More recently, Benkovitz et al. (2020) found that a ratio of 0.6 distinguishes euxinic deposition during sapropel deposition in Holocene sediments from the Eastern Mediterranean Sea, suggesting that the equivocal Fe_{py}/Fe_{HR} zone should be extended further (Figure 1).

In the case of Eastern Mediterranean sapropel deposition, lower Fe_{py}/Fe_{HR} ratios during euxinic deposition likely arise due to the high concentration and highly crystalline nature of Fe (oxyhydr)oxide minerals supplied to the region by the Nile river (see Poulton and Raiswell, 2002), the latter resulting in relatively slow reaction kinetics with respect to Fe (oxyhydr)oxide sulfidation (e.g., Poulton et al., 2004a). Other factors that may influence the extent of sulfidation of Fe_{HR} under euxinic conditions include the concentration of dissolved sulfide in the water column and during diagenesis, and the exposure time of Fe (oxyhydr)oxides to dissolved sulfide (see Canfield et al., 1992; Poulton et al., 2004a). For example, in the low-sulfate, euxinic Lake Cadagno, Switzerland, the limited exposure time between Fe (oxyhydr)oxide minerals and dissolved sulfide results in Fe_{py}/Fe_{HR} ratios of ~0.6 during diagenesis (Xiong et al., 2019). Given that low-sulfate euxinia characterised many episodes of ancient water-column anoxia (Xiong et al., 2019), relatively low Fe_{py}/Fe_{HR} ratios may be a common feature of euxinia in the rock record. This clearly highlights the importance of a multi-proxy approach for robust characterisation of ancient water-column euxinia, including petrographic observations of pyrite morphology (see Section 4).

Recently, Algeo and Li (2020) proposed that, wherever possible, all paleoredox proxies should be internally calibrated for each individual paleodepositional system. In terms of the Fe-speciation proxy, Algeo and Li (2020) investigated a parameter termed DOP_T, which essentially refers to the ratio Fe_{py}/Fe_T (note, not Fe_{py}/Fe_{HR}), where Fe_{py} was estimated (stoichiometrically) from the total S content of the sample, assuming that all of the S was present as pyrite (which is unlikely to be the case). While this is not an accepted or widely used paleoredox proxy and its use is not advocated here, as, amongst other limitations, it could never permit identification of anoxic ferruginous conditions, the general premise behind internal paleoredox calibration is very strong. Indeed, the introduction of equivocal zones in the Fe-speciation paleoredox scheme almost a decade ago was, at least partly, aimed at acknowledging this fact. Nevertheless, despite the introduction of this degree of caution, the boundary values for Fe speciation should be considered indicative, rather than prescriptive, and an internally consistent evaluation of redox conditions based on either a local calibration (if possible) or a multi-proxy approach is warranted. This may be particularly significant in the aforementioned case where sediments deposited under anoxic non-sulfidic conditions may actually constitute a source of Fe^{2+} to the oceans rather than being a sink for Fe_{HR} (see Chen et al., 2020). In this scenario, a 'false' oxic signal may arise from loss of Fe_{HR} to the overlying water column, which may potentially be resolved by additionally considering the co-behaviour of redox-sensitive trace metals.

3 Methodology

3.1 Overview

The sequential extraction scheme of Poulton and Canfield (2005) was essentially designed as a 'one size fits all' approach to quantifying different Fe pools in recent sediments (which refers to pre-lithified sediments) and ancient rocks. It was not necessarily meant to be utilised in its entirety for an individual application. Rather, different aspects of the scheme were intended to be tailored to the nature of the material being analysed and the particular requirement of an individual study. Unfortunately, this point has not always been appreciated, and in some cases this has led to inappropriate usage of the extraction scheme for the required purpose. In addition, it has become apparent that for recent sediments, a modified extraction scheme is often preferable (e.g., Zegeye et al., 2012; Xiong et al., 2019). As such, revised details of the extraction protocol are outlined in the following sub-sections for specific application to recent sediments and ancient rocks.

For both recent sediments and ancient rocks, in addition to the sequential stages detailed in the following sub-sections, a number of additional extractions are performed on separate sub-samples. In all cases, samples are subjected to the two-step chromous chloride distillation procedure of Canfield et al. (1986). The first step involves the dissolution of acid-volatile sulfides, which are dissolved with near-boiling 6 N HCl, whereas the second step uses chromous chloride to release the sulfide present as pyrite. If present, Fe_{AVS} is also extracted in the first sequential step used for recent sediments, and hence the Fe_{AVS} determined by the Canfield et al. (1986) technique should be subtracted from this fraction (Poulton and Canfield, 2005). The situation is more complex when Fe_{AVS} is present in ancient rocks, as it tends to be in a more resistant form. Here, separate checks should be made to ensure no Fe_{AVS} remains after the sequential extraction steps, and Fe_{AVS} should then be subtracted from the total unsulfidised Fe_{HR} pool, as determined by the sum of the sequential steps (Poulton et al., 2010).

A boiling, concentrated HCl extraction may also be used (Raiswell et al., 1988; 1994) if determination of Fe_{PRS} is required (see discussion later). This extraction is widely performed incorrectly, which is particularly problematic since it is highly operationally specific. The sample is gradually brought to the boil in a glass test tube over the course of 60 seconds and is then vigorously boiled for a further 60 seconds, followed by immediate quenching with deionised water. Other variations on this extraction are not acceptable and should be avoided, as they do not produce consistent results in terms of calibrated thresholds (see Raiswell et al., 1994). The boiling HCl extraction may be performed either sequentially, after completion of the initial steps outlined in the following sub-sections 3.2 and 3.3, or on a separate sub-sample. In the latter case, Fe_{PRS} is determined by subtracting the sum of the first three sequential steps plus Fe_{AVS} from the Fe dissolved via the boiling HCl extraction. Finally, a measurement of total Fe (Fe_T) is required for all samples, and this may be achieved via any standard, well-calibrated technique.

3.2 Sequential Protocol for Recent Sediments

a) Samples are subjected to a 0.5 N HCl extraction for 1 h. This extraction has the advantage that both Fe(II) and Fe(III) concentrations can be measured. Fe(II) is measured immediately via spectrophotometer (Stookey, 1970), and the phases targeted include surface-reduced Fe(II) and a variety of other easily extractable reduced minerals, such as poorly crystalline Fe(II) carbonate, Fe(II) phosphate and Fe_{AVS}. Subtraction of Fe_{AVS} determined via the Canfield et al. (1986) technique gives Fe(II) in unsulfidised phases ($Fe(II)_{unsulf}$). The total Fe concentration dissolved by this extraction can be measured via any standard technique, and subtraction of $Fe(II)_{unsulf}$ targets easily reducible ferric minerals, such as ferrihydrite (Fe_{ox1}).

b) The sample is then treated with a sodium dithionite solution (50 g L^{-1} sodium dithionite, 58.82 g L^{-1} tri-sodium citrate, 20 ml L^{-1} acetic acid) for 2 h. The sodium dithionite solution should always be prepared immediately prior to use in order to avoid oxidation of the solution and hence a lower extraction potential. This step (Fe_{ox2}) primarily targets crystalline ferric (oxyhydr)oxide minerals.

c) The final step utilises a 0.2 M ammonium oxalate/0.17 M oxalic acid solution with a treatment time of 6 h. This step targets mixed ferric/ferrous oxides, such as magnetite (Fe_{mag}).

3.3 Sequential Protocol for Ancient Rocks

a) Samples are subjected to a 1 M sodium acetate solution buffered with acetic acid (pH = 4.5) for 48 h at a constant temperature of 50°C. This step primarily targets iron associated with carbonate phases (Fe_{carb}).
b) Samples are then treated with sodium dithionite as described in the previous protocol. This step primarily targets ferric (oxyhydr)oxide minerals (Fe_{ox}).
c) The final step involves treatment with an ammonium oxalate solution as described in the previous protocol. This step targets mixed ferric/ferrous oxides, such as magnetite (Fe_{mag}).

3.4 Further Considerations

More specific analytical details are given in Alcott et al. (2020), which includes the development of a series of international Fe-speciation reference materials for paleoredox analysis. There is growing concern that as Fe speciation has increasingly become a widely used geochemical technique, unverified operational modifications made by individual laboratories may be introducing significant variability in the quality of Fe-speciation analyses. It is stressed here that, for clear reasons of appropriate interpretational utility, analyses should be conducted in the same manner as in the laboratories that developed, tested and calibrated the Fe-speciation method. While an inter-laboratory calibration of a reference material suitable for recent sediments has not yet been developed, the reference materials developed by Alcott et al. (2020) for Fe-speciation analyses of ancient rocks should be used as an essential validation of in-house operational procedures, as is common with a wide variety of standard geochemical and isotopic techniques.

It is also stressed here that, while the extraction scheme of Poulton and Canfield (2005) targets broad mineral groups, it was not intended to be mineral specific. Neither was the scheme, or the concept of highly reactive Fe, developed as a means to quantify the reactivity of particulate Fe towards microbial reduction (c.f., Laufer et al., 2020). As outlined earlier, the extractions are operationally defined, and the Fe_{HR} pool is defined based on its reactivity towards dissolved sulfide. The precise minerals extracted in each sequential step (and the extent of their dissolution) will vary dependent on mineral crystallinity (see Raiswell et al., 1994) and several other factors, including impurities within the mineral structure (see Poulton et al., 2004a). Indeed, Slotznick et al. (2020) highlight discrepancies between 'target' mineral phases and the actual minerals dissolved in each extraction step. However, the Fe dissolved in each extraction step can be considered to comprise an iron pool of similar reactivity towards dissolved sulfide, and it is this factor that has been calibrated in the use of Fe speciation as a paleoredox indicator. For example, since some particularly reactive clay minerals, such as nontronite and glauconite, are partially extracted during the early steps of the Poulton and Canfield (2005) extraction scheme (Raiswell et al., 1994; Hepburn et al., 2020), the Fe extracted from such minerals is, by definition, highly reactive. Hence, as recognised by Hepburn et al. (2020), the calibrated Fe-speciation paleoredox proxy remains robust regardless of the precise minerals extracted in each step, as it is predicated on the overall quantification of Fe_{HR} in marine sediments and not individual minerals.

4 Case Studies

4.1 Recognising Diagenetic Loss of Fe_{HR}

As outlined earlier, loss of unsulfidised Fe_{HR} due to clay mineral formation during diagenesis and metamorphism can significantly diminish primary depositional Fe_{HR}/Fe_T ratios. This process can be recognised via consideration of two additional parameters, Fe_T/Al and Fe_{PRS}/Fe_T, and, when combined, these additional parameters may provide a particularly robust evaluation of potential Fe_{HR} loss. This case study focuses on ~1.88 to 1.83 Ga drill core from the Animikie group, Superior region, North America (Poulton et al., 2004b; 2010). In Figure 2, four drill core sections are shown, progressively deepening in terms of water depth from top to bottom. The red symbols, representing iron formation samples from the Gunflint (shallowest core) and the equivalent deeper water Biwabik (bottom three cores) formations, show clear enrichments in Fe_{HR}/Fe_T coupled with low Fe_{py}/Fe_{HR} ratios, which is entirely consistent with deposition under ferruginous conditions. The shallowest core then transitions into shales of the Rove Formation, and, while Fe_{HR}/Fe_T ratios (black circles) show some variability, there are abundant samples with elevated Fe_{HR}/Fe_T, suggestive of continued water-column anoxia (see Poulton et al., 2004b). For these samples, Fe_{py}/Fe_{HR} ratios initially support a transitional phase between ferruginous and euxinic conditions before the development of more persistent euxinia further up the succession, as indicated by high Fe_{py}/Fe_{HR} ratios (Poulton et al., 2004b). However, while some of the Virginia Formation shales that overlie the Biwabik Iron Formation samples in the three deeper cores have elevated Fe_{HR}/Fe_T ratios, many samples fall in the equivocal zone (Figure 2).

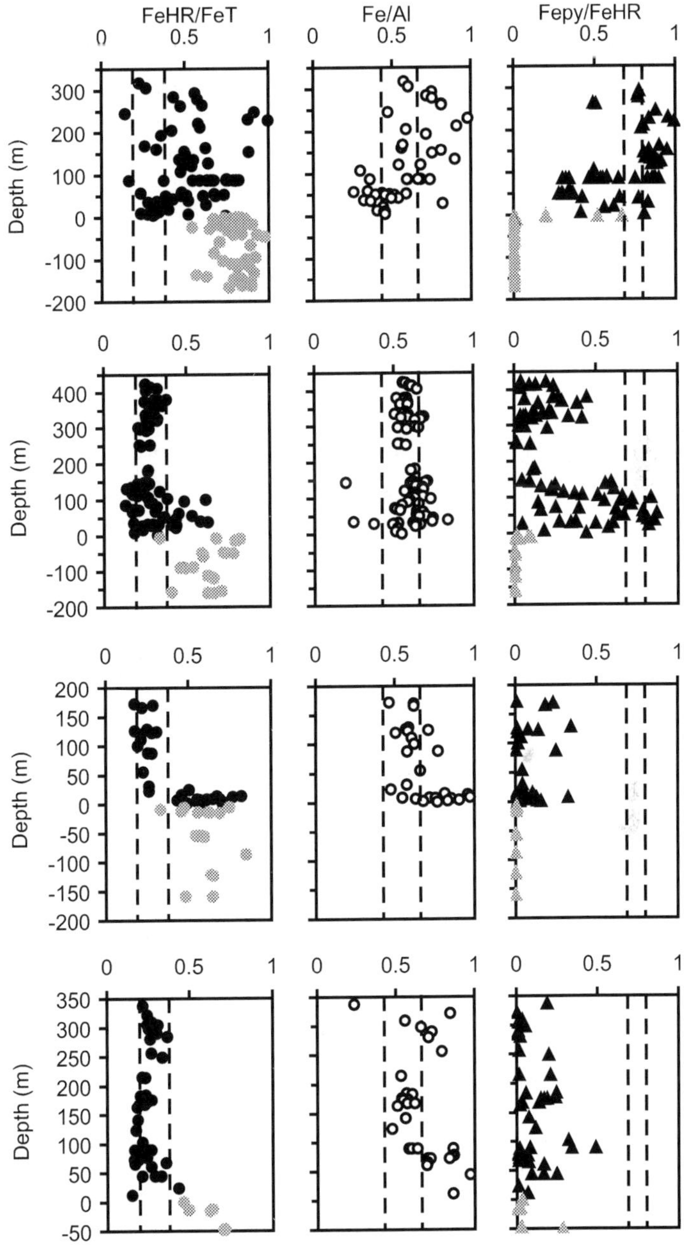

Figure 2 Geochemistry of ~1.88–1.83 Ga drill cores from the Animikie basin, North America (modified after Poulton et al., 2010). Uppermost core represents the shallowest water depth, with progressive deepening of the water column in the cores below. Red symbols represent banded iron formation samples of the Gunflint (shallowest core) and Biwabik (deeper cores) iron formations; black symbols represent shales of the overlying Rove Formation (shallowest core) and Virginia Formation (deeper cores).

To assess the depositional redox history of the equivocal Virginia Formation samples, we initially consider Fe_T/Al ratios, since this ratio is not affected by re-partitioning of Fe during diagenesis (Figure 2). Some of these samples fall above the upper boundary that tends to characterise oxic depositional Fe_T/Al ratios (0.44–0.66; Clarkson et al., 2014), suggesting deposition from anoxic waters. However, many of the samples continue to fall in the Fe_T/Al range characteristic of oxic depositional conditions. Here, it is important to consider the local geochemical characteristics of the sediments. It is clear from many of the non-euxinic samples in the lower Rove Formation of the shallowest core that background Fe_T/Al ratios in these rocks are low, with values that commonly fall below the lower boundary for average shale. Therefore, there is a distinct possibility that low Fe_T/Al ratios in the deeper cores may be due to a low background Fe_T/Al ratio in terrestrial sediments supplied to this region, rather than to deposition under oxic conditions. Thus, to consider the possibility of Fe_{HR} loss to clay minerals, we next consider the Fe_{PRS} content of the three deeper cores.

Common clay minerals that form during diagenesis are extensively dissolved as part of the Fe_{PRS} pool (Raiswell et al., 1994). Fe_{PRS}/Fe_T ratios for these Virginia Formation siliciclastics (0.51 ± 0.08) are considerably elevated relative to Phanerozoic sediments deposited from oxic bottom waters (0.38 ± 0.09; Poulton and Raiswell, 2002), which is also the case in terms of absolute concentrations (see Poulton et al., 2010). Together, these observations are entirely consistent with extensive loss of unsulfidised Fe_{HR} to clay minerals, suggesting that the 'equivocal' Fe_{HR}/Fe_T samples of the Virginia Formation were dominantly deposited under anoxic ferruginous conditions (Poulton et al., 2010). This approach may be further extended to allow estimation of original depositional Fe_{HR}/Fe_T ratios by applying a correction factor that addresses loss of Fe_{HR} to Fe_{PRS} (see Doyle et al., 2018).

4.2 Distinguishing Euxinia from Intense Diagenetic Sulfide Production

For a variety of reasons, including the dominance of low-sulfate conditions during many ancient periods of anoxia (see earlier), it is common for ferruginous depositional settings to record low Fe_{py}/Fe_{HR} ratios (e.g., Poulton et al., 2004b; 2010). However, extensive sulfide production during organic-rich marine sediment diagenesis will progressively increase Fe_{py}/Fe_{HR} ratios, potentially giving a false euxinic signal under ferruginous conditions (such confusion should not occur during oxic deposition, since these settings should be characterised by low Fe_{HR}/Fe_T ratios). The differential behaviour of redox-sensitive trace metals as sulfide becomes more prevalent in the water column provides a potential means to support Fe-speciation-derived inferences on redox state (e.g., Algeo and

Tribovillard, 2009). However, pyrite morphology may also provide important insight, and, for Phanerozoic marine sediments, the use of pyrite framboid size range has been widely used as a means to recognise water column euxinia, whereby small framboids with a narrow range in diameter are commonly considered diagnostic of sulfide in the water column (e.g., Wignall and Newton, 1998). Strict application of the pyrite framboid technique is, however, problematic (see Rickard, 2019), and furthermore, for reasons that are not fully understood, but which may relate to differences in oxidative pathways (framboids tend to form via an intermediate partially oxidised greigite phase; e.g., Wilkin and Barnes, 1997), framboids are highly uncommon in older Precambrian rocks. Nevertheless, as explored here, pyrite morphology may still provide important insight to aid distinction between water column euxinia and intense diagenetic sulfide production during diagenesis beneath a ferruginous water column.

This case study focuses on ~2.65–2.48 Ga drill core samples from the Campbellrand Subgroup, Transvaal Supergroup, South Africa (Schröder et al., 2006). The samples display a complex depositional redox history (Figure 3), several facets of which have been explored elsewhere (Kendall et al., 2010; Zerkle et al., 2012). However, based on a combination of either elevated Fe_{HR}/Fe_T or elevated Fe_T/Al, a broad distinction between oxic and anoxic depositional conditions can be achieved (Figure 3). There are several intervals where highly elevated Fe_{HR}/Fe_T or Fe_T/Al ratios coincide with high Fe_{py}/Fe_{HR}. Here, we focus on the morphology of pyrite in two prominent potentially euxinic horizons – in the Monteville Formation and in the younger Klein Naute Formation (Figure 3).

In the case of the older Monteville Formation samples, reflected light microscopy shows pyritic wavy layers, rip up clasts and nodules, and poorly disseminated coarser-grained euhedral pyrite (Figure 3). These are not typical textures observed in sediments deposited beneath a euxinic water column, where, instead, fine-grained dispersed pyrite tends to dominate (Wignall and Newton, 1998). Thus, the high Fe_{HR}/Fe_T and Fe_{py}/Fe_{HR} ratios of these samples most likely reflect intense diagenetic pyrite formation close to the sediment–water interface beneath a dominantly ferruginous water column. By contrast, pyrite in the lower part of the younger Klein Naute Formation consists of fine-grained, well-disseminated pyrite with only occasional larger euhedral grains, likely formed during diagenesis. Indeed, the pyrite morphology of these samples has been likened to relic framboids (Farquhar et al., 2013), and their small size and limited size range are consistent with morphologies found in sediments deposited under euxinic conditions (Wignall and Newton, 1998). Thus, despite elevated Fe_{HR}/Fe_T and Fe_{py}/Fe_{HR} ratios in both portions of the succession, only the upper Klein Naute Formation likely records water-column euxinia.

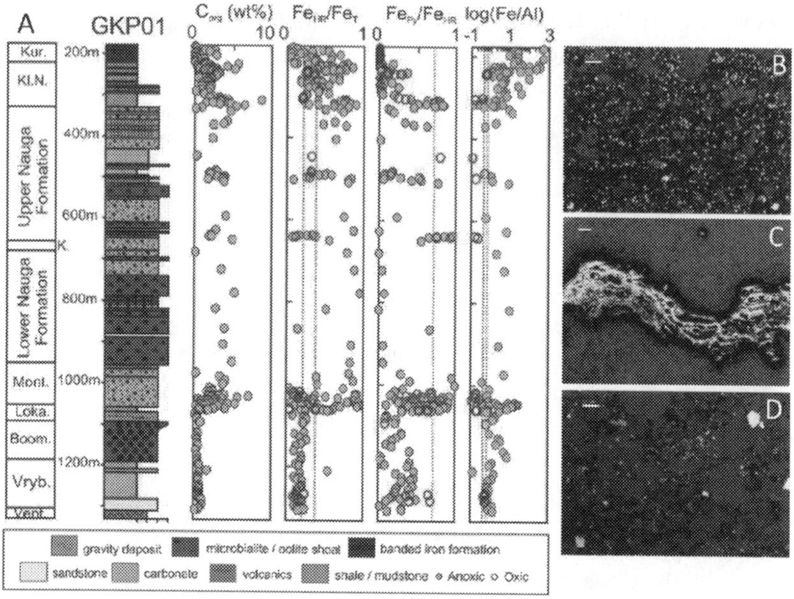

Figure 3 Geochemistry of ~2.65–2.48 Ga drill core from the Campbellrand Subgroup, Transvaal Supergroup, South Africa. A. Organic carbon concentrations (C_{org}) and Fe_{HR}/Fe_T, Fe_{py}/Fe_{HR} and Fe_T/Al ratios; B. Representative reflected light image of finely disseminated pyrite from a 'euxinic' portion of the Klein Naute Formation; C. Representative reflected light image of a high-Fe_{py}/Fe_{HR}, layered pyrite sample from the Monteville Formation; D. Representative reflected light image of poorly dispersed, variable-grain-size pyrite from a high-Fe_{py}/Fe_{HR} sample from the Monteville Formation. Stratigraphic column modified after Schröder et al. (2006). Geochemical data and reflected light images from Kendall et al. (2010).

White scale bars on B, C and D represent 50 μm. Kur. – Kuruman Iron Formation; Kl. Na. – Klein Naute Formation; K. – Kamden Member; Mont. – Monteville Formation; Loka. – Lokamonna Formation; Boom. – Boompaas Formation; Vryb. – Vryburg Formation; Vent. – Ventersdorp Group.

4.3 Benthic Fauna and the Potential Impact of Bioturbation

One factor that may affect interpretation of Fe_{HR}/Fe_T ratios concerns intermittent anoxia and the role of bioturbation during oxic interludes. The degree to which bioturbation affects Fe_{HR}/Fe_T ratios will depend on a variety of factors, including the intensity (particularly depth) of bioturbation, the frequency of oxic versus anoxic conditions (e.g., seasonal versus orbitally paced), and the magnitude of the anoxic water-column Fe_{HR} depositional flux in the particular setting of interest. Thus, intermittent anoxia accompanied by bioturbation during oxic intervals could potentially result in a wide range of Fe_{HR}/Fe_T ratios, giving anything from an oxic to an anoxic signal, dependent on the precise interplay of these parameters.

Here, we first consider potential inconsistency between paleoredox indicators and the presence of benthic fossils in a setting where sediment lamination has not been completely destroyed by bioturbation. A prevalent example concerns the late Cambrian (~501–494 Ma) Alum Shale of Scandinavia. Gill et al. (2011) utilised Fe speciation and a variety of additional redox proxies to demonstrate widespread euxinia in the later Cambrian ocean, including during deposition of the Alum Shale. However, the Alum Shale also contains abundant benthic fossils, which would commonly be considered inconsistent with water-column euxinia. In a particularly well-executed study, Dahl et al. (2019) resolved this apparent paradox by utilising a high-resolution geochemical approach to demonstrate brief oxic interludes of 600–3000 years' duration (corresponding to 1–5 mm stratigraphic thickness), during which opportunistic species were able to recolonise the seafloor (see also Sperling et al., 2016). Thus, far from giving a 'false' euxinic signal, the Fe-speciation data of Gill et al. (2011) actually provide a time-integrated signal of the dominant redox state over the interval of sample thickness.

While this example demonstrates how inconsistencies between the geochemical and paleontological record may potentially be resolved, in more heavily bioturbated sections, the sediment may be too disturbed to preserve any distinction between oxic and anoxic intervals, even at very high resolution. Thus, overall elevated Fe_{HR}/Fe_T ratios in bioturbated successions may be entirely consistent with intermittent intervals of anoxia. Similarly, low Fe_{HR}/Fe_T ratios in such sections do not necessarily preclude intervals of anoxia, since anoxic Fe_{HR}/Fe_T enrichments may be masked by mixing with sediment deposited under oxic conditions. These observations highlight the difficulties inherent in applying Fe speciation to bioturbated sediments, which may ultimately be difficult to resolve, even with a detailed high-resolution geochemical and sedimentological approach. However, the presence of moderate enrichments in both Fe-speciation ratios and redox-sensitive trace metals may provide some insight into average depositional redox conditions when interpreted with care.

5 Outlook

In this review, several facets of the Fe-speciation paleoredox proxy have been outlined, including details of the correct methodological approach and potential complications when interpreting Fe-speciation data, as well as some common misunderstandings of the proxy. As is the case with every standard, widely used geochemical or isotopic proxy, the Fe-speciation paleoredox proxy has come under considerable scrutiny and is also often misused. However, Fe speciation remains the most well-understood and well-calibrated paleoredox tool, and due to its development over time, particularly in terms of potential limitations and the concept of equivocal interpretational zones where additional insight is considered essential, it remains a particularly robust proxy. Indeed, when used in an appropriate manner, Fe speciation can be a remarkable tool, which not only provides a level of detailed insight into local redox chemistry that is unachievable by any other paleoredox proxy, but also forms a framework for the wider consideration and interpretation of a variety of additional geochemical tools. In this regard, Fe speciation often forms an essential basis for subsequent geochemical and isotopic evaluation of regional to global redox characteristics (e.g., Gill et al., 2011; Kendall et al., 2011) as well as for understanding redox-driven nutrient dynamics (e.g., Bowyer et al., 2020; Guilbaud et al., 2020).

As highlighted throughout this review, although Fe speciation is a primary paleoredox tool, reconstructions are best achieved when tackled through a multi-proxy approach. This can be challenging, particularly in the common case where different paleoredox proxies, or contrasting observations amongst geochemical, paleontological or sedimentological parameters, may at first sight appear to provide contradictory information. As highlighted earlier in the case of opportunistic colonisation by benthic fauna, such complications may be resolved by deeper thought and investigation rather than by simply dismissing certain facets of the data, or the data of one particular redox proxy. In all cases, an internally consistent interpretation that satisfies all observations should be the aim, and when this is achieved, a more realistic reconstruction of paleoenvironmental conditions is undoubtedly attained.

6 Key Papers

Methodological

Alcott, L. J., Krause, A. J., Hammarlund, E. U., et al. (2020). Development of iron speciation reference materials for paleoredox analysis. *Geostandards and Geoanalytical Research*, 44, 581–591. *Reports methodological advances and develops, via an inter-laboratory calibration, a series of international reference materials for iron speciation.*

Canfield, D. E., Raiswell, R., Westrich, J. T., et al. (1986). The use of chromium reduction in the analysis of reduced inorganic sulfur in sediments and shales. *Chemical Geology*, 54, 149–155. *Development of a two-step protocol for determining the concentrations of acid volatile sulfide and pyrite in sediments and rocks.*

Poulton, S. W., Canfield, D. E. (2005). Development of a sequential extraction procedure for iron: implications for iron partitioning in continentally derived particulates. *Chemical Geology*, 214, 209–221. *Reports the development of a sequential extraction scheme for fully evaluating highly reactive iron in natural sediments. Includes additional extraction steps aimed at quantifying iron not extracted by existing schemes, including carbonate phases and magnetite. Has become the standard technique in the use of iron speciation as a paleoredox indicator.*

Raiswell, R., Canfield, D. E., Berner, R. A. (1994). A comparison of iron extraction methods for the determination of degree of pyritization and the recognition of iron-limited pyrite formation. *Chemical Geology*, 111, 101–110. *Tests a variety of extraction techniques for determining concentrations of iron in several common synthetic and natural iron-bearing minerals. Calibrates the use of sodium dithionite for determining non-sulfidised highly reactive iron phases in sediments.*

Developmental

Canfield, D. E. (1989). Reactive iron in marine sediments. *Geochimica et Cosmochimica Acta*, 53, 619–632. *A seminal paper that explores the role of iron (oxyhydr)oxide mineral reactivity, and both dissimilatory iron reduction and abiotic reduction of iron (oxyhydr)oxide minerals by dissolved sulfide, in controlling pore-water profiles and solid-phase iron partitioning in marine sediments.*

Canfield, D. E., Lyons, T. W., Raiswell, R. (1996). A model for iron deposition to euxinic Black Sea sediments. *American Journal of Science*, 296, 818–834. *Proposes a model to explain enrichments in sulfidised iron in sediments deposited beneath euxinic waters in the Black Sea, whereby water column sulfidation enhances the flux of sulfidised highly reactive iron to the sea floor. The first paper to provide a mechanistic understanding of iron enrichments, thereby allowing the development of iron speciation as an indicator of water column euxinia.*

Canfield, D. E., Raiswell, R., Bottrell, S. H. (1992). The reactivity of sedimentary iron minerals toward sulfide. *American Journal of Science*, 292, 659–683. *Compiles existing and new data on the reactivity of a variety of synthetic and natural iron minerals towards dissolved sulfide. Demonstrates large differences in relative mineral reactivity and uses this insight to evaluate controls on pyrite formation in marine sediments.*

Clarkson, M. O., Poulton, S. W., Guilbaud, R., Wood, R. A., 2014. Assessing the utility of Fe/Al and Fe-speciation to record water column redox conditions in carbonate-rich sediments. *Chemical Geology*, 382, 111–122. *Provides an updated calibration of iron speciation in a variety of modern and ancient marine sediments. Specifically evaluates the potential use, and limitations, of applying iron speciation to carbonate-rich lithologies.*

Poulton, S. W., Canfield, D. E. (2011). Ferruginous conditions: A dominant feature of the ocean through Earth's history. *Elements*, 7, 107–112. *Summarises how iron speciation revolutionised understanding of the significance of ferruginous water-column conditions through Earth's history. Specifically refines boundary conditions for the use of iron speciation, with the introduction of 'equivocal' zones in recognition of the potential for misclassification when applying a strict interpretation of previously defined boundary conditions.*

Poulton, S. W., Fralick, P. W., Canfield, D. E, (2004). The transition to a sulphidic ocean ~1.84 billion years ago. *Nature*, 173–177. *The first paper to utilise the updated extraction scheme of* Poulton and Canfield (2005) *to reconstruct ancient water-column redox conditions. Introduces the concept of using the extent of sulfidation of highly reactive iron, alongside ratios of highly reactive iron to total iron, to provide a new method for distinguishing oxic, ferruginous and euxinic water-column conditions.*

Poulton, S. W., Krom, M. D., Raiswell, R. (2004). A revised scheme for the reactivity of iron (oxyhydr)oxide minerals towards dissolved sulfide. *Geochimica et Cosmochimica Acta*, 68, 3703–3715. *Provides the first internally consistent evaluation of the reaction kinetics of the key iron (oxyhydr)oxide minerals towards dissolved sulfide. Demonstrates that iron (oxyhydr)oxide minerals fall into two distinct reactivity groups, with haematite and magnetite reacting at a slower rate than minerals such as ferrihydrite and lepidocrocite.*

Poulton, S. W., Raiswell, R. (2002). The low-temperature geochemical cycle of iron: From continental fluxes to marine sediment deposition. *American Journal of Science*, 302, 774–805. *Provides the first overview of the modern global iron cycle based on iron speciation, including weathering controls on the magnitude of highly reactive iron delivered to marine sediments, and the significance of inner shore settings for trapping riverine highly reactive iron. Proposes a long-term steady state for the global iron cycle through the Phanerozoic.*

Raiswell, R., Canfield, D. E. (1996). Rates of reaction between silicate iron and dissolved sulfide in Peru Margin sediments. *Geochimica et Cosmochimica Acta*, 60, 2777–2787. *Uses natural sediments to specifically estimate the reactivity of poorly reactive iron silicates towards dissolved sulfide. Demonstrates that such phases are clearly distinct from iron (oxyhydr)oxide minerals, in that they are only appreciably reactive on a million-year timescale.*

Raiswell, R., Canfield, D. E. (1998). Sources of iron for pyrite formation in marine sediments. *American Journal of Science*, 298, 219–245. *Provides a global compilation of the unsulfidised and sulfidised highly reactive iron content of marine sediments from a variety of water-column redox settings. Establishes the widespread utility of iron speciation as a tool for identifying euxinic water-column conditions and promotes distinct boundary conditions for this identification relative to oxic/dysoxic settings.*

Raiswell, R., Newton, R., Wignall, P. B. (2001). An indicator of water-column anoxia: Resolution of biofacies variations in the Kimmeridge Clay (Upper Jurassic, U.K.). *Journal of Sedimentary Research*, 71, 286–294. *Provides the first calibration of highly reactive iron as a paleoredox tool for ancient sediments and defines this as an 'indicator of anoxicity'. The iron-speciation proxy is calibrated relative to different oxygen-restricted biofacies, as defined by the nature and species richness of the benthic fauna present.*

Reviews

Lyons, T. W., Severmann, S. (2006). A critical look at iron paleoredox proxies: New insights from modern euxinic marine basins. *Geochimica et Cosmochimica Acta*, 70, 5698–5722. *A scholarly review of the use of iron speciation and Fe/Al ratios for identifying water-column euxinia. Specifically highlights mechanistic issues and the significance of the shelf-to-basin shuttle for explaining iron enrichments in proximal euxinic settings.*

Raiswell, R., Canfield, D. E. (2012). The iron biogeochemical cycle past and present. *Geochemical Perspectives*, 1, 1–232. *A comprehensive review of the global biogeochemical iron cycle. A must read for scholars of iron biogeochemistry, summarising several decades of innovative science.*

Raiswell, R., Hardisty, D. S., Lyons, et al. (2018). The iron paleoredox proxies: A guide to the pitfalls, problems and proper practice. *American Journal of Science*, 318, 491–526. *A detailed, up-to-date review of proper practice in the application of iron speciation as a paleoredox indicator. Includes several case studies highlighting specific complications when employing iron speciation. Specifically highlights the value of iron speciation as a local redox indicator that, when combined with redox-sensitive trace metal proxies or isotopic proxies, can shed light on the global redox state of the oceans through time.*

References

Alcott, L. J., Krause, A. J., Hammarlund, E. U. (2020) Development of iron speciation reference materials for paleoredox analysis. *Geostandards and Geoanalytical Research*, 44, 581–591.

Algeo, T. J., Tribovillard, N. (2009) Environmental analysis of paleoceanographic systems based on molybdenum-uranium covariation. *Chemical Geology*, 268, 211–225.

Algeo, T. J., Li, C. (2020) Redox classification and calibration of redox thresholds in sedimentary systems. *Geochimica et Cosmochimica Acta*, 287, 8–26.

Anderson, T. F., Raiswell, R. (2004) Sources and mechanisms for the enrichment of highly reactive iron in euxinic Black Sea sediments. *American Journal of Science*, 304, 203–233.

Benkovitz, A., Matthews, A., Teutsch, N., et al. (2020) Tracing water column euxinia in Eastern Mediterranean Sapropels S5 and S7. *Chemical Geology*, 545, 119627.

Berner, R. A. (1970) Sedimentary pyrite formation. *American Journal of Science*, 268, 1–23.

Berner, R. A. (1984) Sedimentary pyrite formation: An update. *Geochimica et Cosmochimica Acta*, 48, 605–615.

Bowyer, F. T., Shore, A. J., Wood, R. A., et al. (2020) Regional nutrient decrease drove redox stabilisation and metazoan diversification in the late Ediacaran Nama Group, Namibia. *Scientific Reports*, 10, 2240.

Canfield, D. E. (1989) Reactive iron in marine sediments. *Geochimica et Cosmochimica Acta*, 53, 619–632.

Canfield, D. E., Raiswell, R., Westrich, J. T., et al. (1986). The use of chromium reduction in the analysis of reduced inorganic sulfur in sediments and shales. *Chemical Geology*, 54, 149–155.

Canfield, D. E., Raiswell, R., Bottrell, S. H. (1992) The reactivity of sedimentary iron minerals toward sulfide. *American Journal of Science*, 292, 659–683.

Canfield, D. E., Lyons, T. W., Raiswell, R. (1996). A model for iron deposition to euxinic Black Sea sediments. *American Journal of Science*, 296, 818–834.

Canfield, D. E., Poulton, S. W., Knoll, A. H., et al. (2008) Ferruginous conditions dominated Later Neoproterozoic deep-water chemistry. *Science*, 321, 949–952.

Chen, X., Li, M., Sperling, E. A., et al. (2020) Mesoproterozoic paleo-redox changes during 1500–1400 Ma in the Yanshan Basin, North China. *Precambrian Research*, 347, 105835.

Clarkson, M. O., Poulton, S. W., Guilbaud, R., et al. (2014) Assessing the utility of Fe/Al and Fe speciation to record water column redox conditions in carbonate-rich sediments. *Chemical Geology*, 382, 111–122.

Crowe, S. A., Jones, C., Katsev, S., et al. (2008) Photoferrotrophs thrive in an Archean ocean analogue. *PNAS*, 105, 15, 938–915,943.

Dahl, T. W., Siggaard-Andersen, M.-L., Schovsbo, N. H., et al. (2019) Brief oxygenation events in locally anoxic oceans during the Cambrian solves the animal breathing paradox. *Scientific Reports*, 9, 11669.

Derry, L. A., Jacobsen, S. B. (1990) The chemical evolution of Precambrian seawater: Evidence from REEs in banded iron formations. *Geochimica et Cosmochimica Acta*, 54, 2965–2977.

Doyle, K. A., Poulton, S. W., Newton, R. J., et al. (2018) Shallow water anoxia in the Mesoproterozoic ocean: Evidence from the Bashkir Meganticlinorium, Southern Urals. *Precambrian Research*, 317, 196–210.

Farquhar, J., Cliff, J., Zerkle, A. L., et al. (2013) Pathways for Neoarchean pyrite formation constrained by mass-independent sulfur isotopes. *PNAS*, 110, 17638–17643.

Gill, B. C., Lyons, T. W., Young, S. A., et al. (2011) Geochemical evidence for widespread euxinia in the Later Cambrian ocean. *Nature*, 469, 80–83.

Guilbaud, R., Poulton, S. W., Butterfield, N. J., et al. (2015) A global transition to ferruginous conditions in the early Neoproterozoic oceans. *Nature Geoscience*, 8, 466–470.

Guilbaud, R., Poulton, S. W., Thompson, J., et al. (2020) Widespread oligotrophic conditions in the early Neoproterozoic ocean. *Nature Geoscience*, 13, 296–301.

Halevy, I., Alesker, M., Schuster, E. M., et al. (2017) A key role for green rust in the Precambrian oceans and the genesis of iron formations. *Nature Geoscience*, 10, 135–139.

Hepburn, L. E., Butler, I. B., Boyce, A., et al. (2020) The use of operationally-defined sequential Fe extraction methods for mineralogical applications: A cautionary tale from Mössbauer spectroscopy. *Chemical Geology*, 543, 119584.

Jiang, C. Z., Tosca, N. J. (2019) Fe (II)-carbonate precipitation kinetics and the chemistry of anoxic ferruginous seawater. *Earth and Planetary Science Letters*, 506, 231–242.

Johnston, D. T., Poulton, S. W., Dehler, C., et al. (2010) An emerging picture of Neoproterozoic ocean chemistry: Insights from the Chuar Group, Grand Canyon, USA. *Earth and Planetary Science Letters*, 290, 64–73.

Kendall, B., Reinhard, C. T., Lyons, T. W., et al. (2010) Pervasive oxygenation along late Archaean ocean margins. *Nature Geoscience*, 3, 647–652.

Kendall, B., Gordon, G. W., Poulton, S. W., et al. (2011) Molybdenum isotope constraints on the extent of late Paleoproterozoic ocean euxinia, *Earth and Planetary Science Letters*, 307, 450–460.

Konhauser, K. O., Hamade, T., Raiswell, R., et al. (2002) Could bacteria have formed the Precambrian banded iron formations? *Geology*, 30, 1079–1082.

Laufer, K., Michaud, A. B., Røy, H., et al. (2020) Reactivity of iron minerals in the seabed toward microbial reduction – a comparison of different extraction techniques. *Geomicrobiology Journal*, 37, 170–189.

Li, Y.-L., Konhauser, K. O., Zhai, M. (2017) The formation of primary magnetite in the early Archean oceans. *Earth and Planetary Science Letters*, 466, 103–114.

Lyons, T. W., Severmann, S. (2006) A critical look at iron paleoredox proxies: New insights from modern euxinic marine basins. *Geochimica et Cosmochimica Acta*, 70, 5698–5722.

Lyons, T. W., Werne, J. P., Hollander, D. J., et al. (2003) Contrasting sulfur geochemistry and Fe/Al and Mo/Al ratios across the last oxic-to-anoxic transition in the Cariaco Basin, Venezuela. *Chemical Geology*, 195, 131–157.

März, C., Poulton, S. W., Beckmann, B., et al. (2008), Redox sensitivity of P cycling during marine black shale formation: Dynamics of sulfidic and anoxic, non-sulfidic bottom waters. *Geochimica et Cosmochimica Acta*, 72, 3703–3717.

Milliman, J. D., Meade, R. H. (1983) World-wide delivery of river sediment to the oceans. *Journal of Geology*, 91, 1–21.

Planavsky, N. J., McGoldrick, P., Scott, C. T., et al. (2011) Widespread iron-rich conditions in the mid-Proterozoic ocean. *Nature*, 477, 448–451.

Poulton, S. W., Canfield, D. E. (2005) Development of a sequential extraction procedure for iron: Implications for iron partitioning in continentally derived particulates. *Chemical Geology*, 214, 209–221.

Poulton, S. W., Canfield, D. E. (2011) Ferruginous conditions: A dominant feature of the ocean through Earth's history. *Elements*, 7, 107–112.

Poulton, S. W., Raiswell, R. (2002) The low-temperature geochemical cycle of iron: From continental fluxes to marine sediment deposition. *American Journal of Science*, 302, 774–805.

Poulton, S. W., Krom, M. D., Raiswell, R. (2004a) A revised scheme for the reactivity of iron (oxyhydr)oxide minerals towards dissolved sulfide. *Geochimica et Cosmochimica Acta*, 68, 3703–3715.

Poulton, S. W., Fralick, P. W., Canfield, D. E. (2004b) The transition to a sulphidic ocean ~1.84 billion years ago. *Nature*, 431, 173–177.

Poulton, S. W., Fralick, P. W., Canfield, D. E. (2010) Spatial variability in oceanic redox structure 1.8 billion years ago. *Nature Geoscience*, 3, 486–490.

Raiswell, R., Al-Biatty, H. J. (1989) Depositional and diagenetic C-S-Fe signatures in early Paleozoic normal marine shales. *Geochimica et Cosmochimica Acta*, 53, 1147–1152.

Raiswell, R., Anderson, T. F. (2005) Reactive iron enrichment in sediments deposited beneath euxinic bottom waters: Constraints on supply by shelf recycling. In: McDonald, I., Boyce, A. J. , Butler, I. B. , et al. (eds.), *Mineral Deposits and Earth Evolution*, Geological Society Special Publication 248, pp. 179–194.

Raiswell, R., Canfield, D. E. (1996) Rates of reaction between silicate iron and dissolved sulfide in Peru Margin sediments. *Geochimica et Cosmochimica Acta*, 60, 2777–2787.

Raiswell, R., Canfield, D. E. (1998) Sources of iron for pyrite formation in marine sediments. *American Journal of Science*, 298, 219–245.

Raiswell, R., Canfield, D. E. (2012) The iron biogeochemical cycle past and present. *Geochemical Perspectives*, 1, 1–232.

Raiswell, R., Buckley, F., Berner, R. A., et al. (1988) Degree of pyritization of iron as a paleoenvironmental indicator of bottom-water oxygenation. *Journal of Sedimentary Research*, 58, 812–819.

Raiswell, R., Canfield, D. E., Berner, R. A. (1994) A comparison of iron extraction methods for the determination of degree of pyritization and the recognition of iron-limited pyrite formation. *Chemical Geology*, 111, 101–110.

Raiswell, R., Newton, R. J., Wignall, P. B. (2001) An indicator of water column anoxia: Resolution of biofacies variations in the Kimmeridge Clay (Upper Jurassic, U.K.). *Journal of Sedimentary Research*, 71, 286–294.

Raiswell, R., Hardisty, D., Lyons, T. W., et al. (2018) The iron paleoredox proxies: A guide to the pitfalls, problems and proper practice. *American Journal of Science*, 318, 491–526.

Rasmussen, B., Krapez, B., Muhling, J. R., et al. (2015) Precipitation of iron silicate nanoparticles in early Precambrian oceans marks Earth's first iron age. *Geology*, 43, 303–306.

Rickard, D. (2019) Sedimentary pyrite framboid size-frequency distributions: A meta-analysis. *Palaeogeography, Palaeoclimatology, Palaeoecology*, 522, 62–75.

Schröder, S., Lacassie, J. P., Beukes, N. J. (2006) Stratigraphic and geochemical framework of the Agouron drill cores, Transvaal Supergroup (Neoarchean–Paleoproterozoic, South Africa). *South African Journal of Geology*, 109, 23–54.

Slotznick, S. P., Sperling, E. A., Tosca, N. J., et al. (2020) Unraveling the mineralogical complexity of sediment iron speciation using sequential extractions. *Geochemistry, Geophysics, Geosystems*, 21, e2019GC008666.

Sperling, E. A., Carbone, C., Strauss, J. V., et al. (2016) Oxygen, facies, and secular controls on the appearance of Cryogenian and Ediacaran body and trace fossils in the Mackenzie Mountains of northwestern Canada. *Geological Society of America, Bulletin*, 128, 558–575.

Stookey, L. L. (1970) Ferrozine – a new spectrophotometric reagent for iron. *Analytical Chemistry*, 42, 779–781.

Tostevin, R., Wood, R. A., Shields, G. A., et al. (2016) Low-oxygen waters limited habitable space for early animals. *Nature Communications*, 7, 12818.

Wignall, P. B., Newton, R. J. (1998) Pyrite framboid diameter as a measure of oxygen deficiency in ancient mudrocks. *American Journal of Science*, 298, 537–552.

Wilkin, R. T., Barnes, H. L. (1997) Formation processes of framboidal pyrite. *Geochimica et Cosmochimica Acta*, 61, 323–339.

Wood, R. A., Poulton, S. W., Prave, A. R., et al. (2015) Dynamic redox conditions control late Ediacaran metazoan ecosystems in the Nama Group, Namibia. *Precambrian Research*, 261, 252–271.

Xiong, Y., Guilbaud, R., Peacock, C. L., et al. (2019) Phosphorus cycling in Lake Cadagno, Switzerland: A low sulfate euxinic ocean analogue. *Geochimica et Cosmochimica Acta*, 251, 116–135.

Zegeye, A., Bonneville, S., Benning, L. G., et al. (2012) Green rust formation controls nutrient availability in a ferruginous water column. *Geology*, 40, 599–602.

Zerkle, A. L., Claire, M. W., Domagal-Goldman, S. D., et al. (2012) A bistable organic-rich atmosphere on the Neoarchaean Earth. *Nature Geoscience*, 5, 359–363.

Zhang, K., Zhu, X., Wood, R. A., et al. (2018) Oxygenation of the Mesoproterozoic ocean and the evolution of complex eukaryotes. *Nature Geoscience*, 11, 345–350.

Acknowledgements

Rob Raiswell, Don Canfield and Tim Lyons, the pioneers of modern Fe-S systematics, are thanked for inspiration and discussion. Rob Raiswell and Erik Sperling are thanked for insightful and helpful reviews. Lewis Alcott is thanked for drafting Figure 3. Support from a Royal Society Wolfson Research Merit Award is gratefully acknowledged.

Cambridge Elements

Elements in Geochemical Tracers in Earth System Science

About the Series

This innovative series provides authoritative, concise overviews of the many novel isotope and elemental systems that can be used as 'proxies' or 'geochemical tracers' to reconstruct past environments over thousands to millions to billions of years—from the evolving chemistry of the atmosphere and oceans to their cause-and-effect relationships with life.

Covering a wide variety of geochemical tracers, the series reviews each method in terms of the geochemical underpinnings, the promises and pitfalls, and the 'state-of-the-art' and future prospects, providing a dynamic reference resource for graduate students, researchers and scientists in geochemistry, astrobiology, paleontology, paleoceanography and paleoclimatology.

The short, timely, broadly accessible papers provide much-needed primers for a wide audience—highlighting the cutting-edge of both new and established proxies as applied to diverse questions about Earth system evolution over wide-ranging time scales.

Cambridge Elements

Elements in Geochemical Tracers in Earth System Science

Elements in the Series

The Uranium Isotope Paleoredox Proxy
Kimberly V. Lau, Stephen J. Romaniello and Feifei Zhang

Triple Oxygen Isotopes
Huiming Bao

Application of Thallium Isotopes
Jeremy D. Owens

Earth History of Oxygen and the iprOxy
Zunli Lu, Wanyi Lu, Rosalind E. M. Rickaby and Ellen Thomas

Selenium Isotope Paleobiogeochemistry
Eva E. Stüeken and Michael A. Kipp

The TEX_{86} Paleotemperature Proxy
Gordon N. Inglis and Jessica E. Tierney

The Pyrite Trace Element Paleo-Ocean Chemistry Proxy
Daniel D. Gregory

Calcium Isotopes
Elizabeth M. Griffith and Matthew S. Fantle

Pelagic Barite: Tracer of Ocean Productivity and a Recorder of Isotopic Compositions of Seawater S, O, Sr, Ca and Ba
Weiqi Yao, Elizabeth Griffith, Adina Paytan

Vanadium Isotopes: A proxy for ocean oxygen variations
Sune G. Neilsen

Cerium Anomalies and Paleoredox
Rosalie Tostevin

Nitrogen Isotopes in Deep Time
Colin Mettam and Aubrey L. Zerkle

The Chromium Isotope System as a Tracer of Ocean and Atmosphere Redox
Kohen W. Bauer, Noah J. Planavsky, Christopher T. Reinhard and Devon B. Cole

The Iron Speciation Paleoredox Proxy
Simon W. Poulton

A full series listing is available at: www.cambridge.org/EESS

Printed in the United States
by Baker & Taylor Publisher Services